LIFTING
BY
LEVERS

ANDREW DUNN

**Illustrated by
ED CARR**

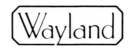

Titles in this series
Heat
It's Electric
Lifting by Levers
The Power of Pressure
Simple Slopes
Wheels at Work

First published in 1991 by
Wayland (Publishers) Ltd
61 Western Road, Hove
East Sussex, BN3 1JD, England

© Copyright 1991 Wayland (Publishers) Ltd

Editor: Anna Girling
Design: Carr Associates Graphics, Brighton

British Library Cataloguing in Publication Data
Dunn, Andrew
    Lifting by levers.–(How things work)
    I. Title   II. Series
    372.3

HARDBACK ISBN 0-7502-0218-1

PAPERBACK ISBN 0-7502-0954-2

Typeset by Dorchester Typesetting Group Ltd
Printed in Italy by G. Canale & C.S.p.A. Turin

Lynburn Primary School
Nith Street
Dunfermline
KY11 4LU

Contents

Words in *italic* in the text are explained in the glossary on page 30.

What is a lever?

First thing in the morning, do you open the bedroom door, or switch on the light? Either way, you are using a lever. Levers can do all sorts of things, and we use them in many different ways. The lever is so very simple it hardly seems to be a *machine*. But that is what it is.

You can see how a machine works just by looking at it, or looking inside. But to understand how a machine works, you must find out about the *principles* behind it.

These tools use levers in many different ways.

A machine is anything which helps us by taking a *force* – the force from your arm, for instance – and transferring it so that it can be used exactly where it is needed. At the same time it makes the force bigger or smaller.

4

Looking for levers

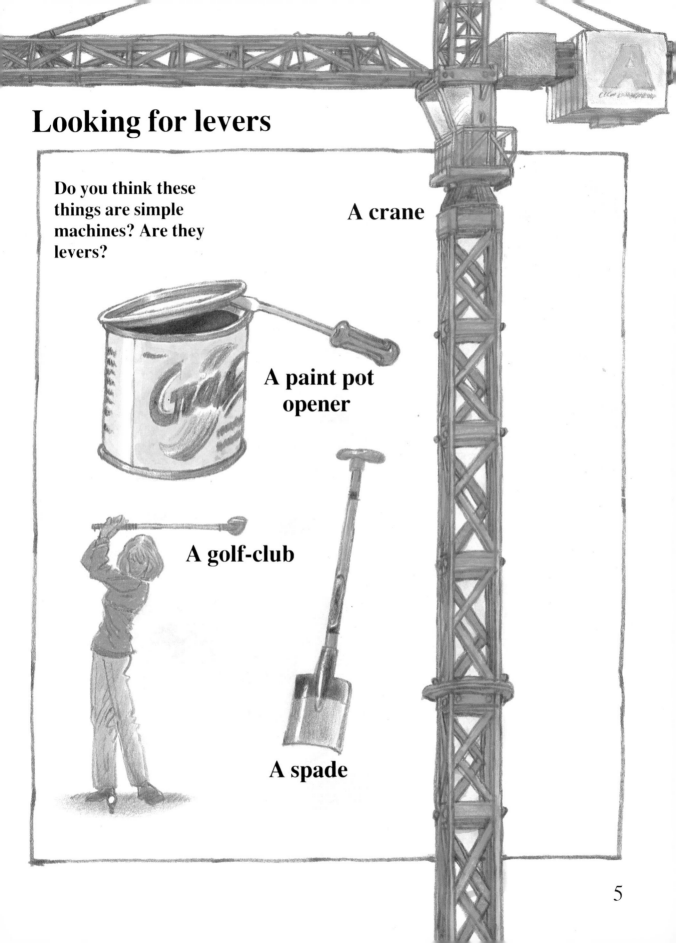

Do you think these things are simple machines? Are they levers?

A crane

A paint pot opener

A golf-club

A spade

5

All in the balance

When did you last play on a see-saw with a friend? You were sitting on a lever!

A see-saw is a plank that tilts up and down on a point, called a *pivot*. When you push down on one end, the other end goes up. If you and your friend are about the same size, and you both sit at the ends of the see-saw, you should be able to *balance* each other. But if you move closer to the centre, your friend will remain firmly on the ground. What has happened? Has your friend got heavier? Have you got lighter?

Before you turn over the page for the answer, try the experiment opposite.

On a see-saw, either child can 'lever' the other off the ground.

You need a pencil, a ruler and a few coins of the same kind.

Put the pencil on a table and balance the ruler on it so that it is like a see-saw. Then put two coins together on the ruler, 2 cm away from the pivot. Put one coin exactly the same distance from the pivot on the other side. What happens?

Now move the single coin slowly away from the pivot. At what point does the ruler balance? Can you see how you have used one coin to lift two?

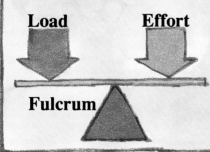

The force you apply to one end of a lever is called the effort. The lever turns on a pivot, or fulcrum, and lifts a load. Lifting a load involves work. The see-saw, a simple lever, has helped you to do more work with less effort!

The lever principle

Now you can see why levers are so useful. You can use less effort to lift the same load, so long as you apply the effort further from the fulcrum. This is because the effort moves through a longer distance than the load. Work is a combination of effort and distance.

The great principle of the lever is: effort times distance from fulcrum equals load times distance from fulcrum.

Any bar or rod pivoting on a fulcrum is a lever. The see-saw is a first-class lever. It is called that, not because it is the best kind of lever, but the simplest. A first-class lever always has its fulcrum between the effort and the load. As you will see later, there are also second and third-class levers.

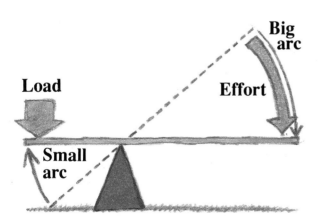

Now look again at your ruler and coins experiment. Not only is the effort twice as far from the fulcrum, but the arc it makes is twice as long as that of the load.

This child's mobile looks simple – but see how different lengths of wire are used to balance the single shapes against the heavier pairs below them.

Throwing some weight around

Can you imagine what would happen if you were sitting on one end of a see-saw, and an enormous rock was dropped on the other end? You would shoot up into the air!

Long ago people used this idea for a war machine called a catapult. They used it to throw rocks or other objects over castle walls.

More first-class levers

Look around you. Can you think of any other first-class levers? Here are some suggestions.

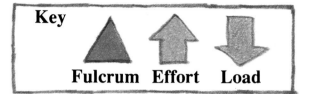

Key

Fulcrum   Effort   Load

Claw hammer

The effort of the hand is *magnified* by the length of the handle. The load is the nail.

Warehouse trolley

The heavy load is very close to the fulcrum. The handles are much further away, making it easy to lift heavy objects and trundle them away.

Pliers

Pliers are a pair of levers *hinged* at the fulcrum. The effort of squeezing the handles together puts a strong grip on the load, which is the resistance of the object being squeezed.

Scissors

Scissors work in the same way as pliers. Here, the *blades* slice through the material as they are squeezed against each other. The load is the resistance of the material being cut.

Spade

The bottom of the spade first slices its way into the soil. Then the soil surface acts as the fulcrum. The effort of the hand is magnified by the length of the handle, so that the digger can lift out a heavy load.

11

The wheelbarrow

A wheelbarrow may not look much like a lever – but it is. It is a second-class lever. Second-class levers always have their load in between the fulcrum and the effort.

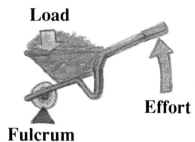

Load

Effort

Fulcrum

If the effort is three times as far from the fulcrum as the load, the effort is only one-third as great as the load.

Second-class levers are just as useful as first-class levers. Can you think of any others, before you turn over the page?

How useful is it?

Do you have this type of wheelbarrow at home? If so, try this experiment. Fill it so that you can just about lift the weight.

Then try to use the legs as the fulcrum instead of the wheel. Press down on the handles as if the wheelbarrow was a first-class lever.

Can you move the load off the ground? Why has it become so difficult?

More second-class levers

These machines are all second-class levers. They are all different but, as you can see, the load is always between the fulcrum and the effort.

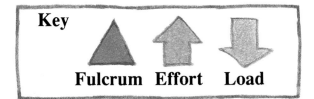
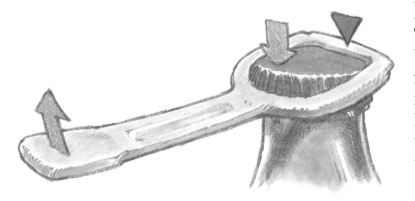

Bottle opener
The far side of the bottle cap is the fulcrum, and the near side is the load. Pulling the handle up levers the cap off the bottle.

Hole punch
The machine used to punch holes in paper magnifies the effort pressing on it, enough to drive the punch through several sheets.

Nutcracker
A nutcracker is a pair of second-class levers hinged at the fulcrum.

What is a door?

Have a look at any door. Does it have a handle or a doorknob? A door handle is a lever. The load is the resistance of the catch *mechanism*.

But look at the door itself. Open it slightly, then try pushing it wide open by pushing the handle away from you. It is quite easy, isn't it?

Now try again, by pushing on the door near its hinges. Is a door a lever? Is it first-class or second-class?

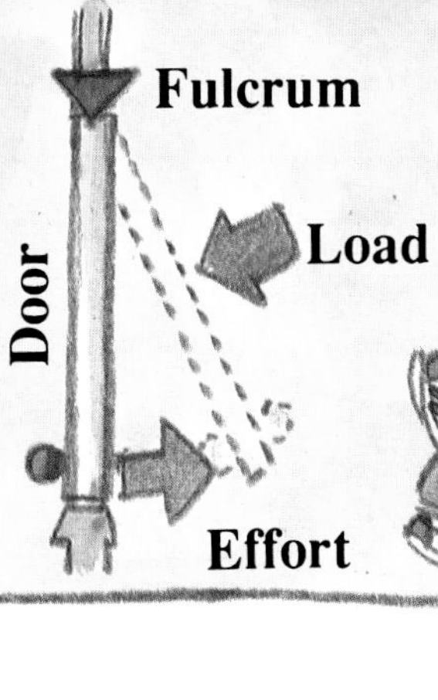

Fulcrum

Load

Door

Effort

15

Another class of levers

A golf-club is a lever. So is a tennis racket, or a cricket bat, or a baseball bat. They are all third-class levers. Like second-class levers, the fulcrum is at one end. But this time the effort is closer to the fulcrum than the load. The effect is that the load moves with less force than the effort, but it moves much further, so it moves much faster.

In a golf swing, the fulcrum is the player's wrist. The head of the club moves much further than the golfer's hands, sending the ball – the load – flying through the air.

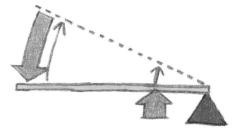

A tennis racket

A golf-club

A baseball bat

A cricket bat

Even your body is made up of levers – in your arms and legs. You do not need a bat to play soccer, do you?

Key

Fulcrum Effort Load

More third-class levers

All these levers are third-class levers. In each one, a small movement at one end produces a bigger movement at the other end.

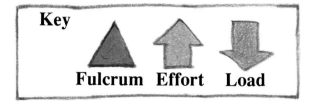

Key

Fulcrum Effort Load

Spade

A spade is a third-class lever when it is used for shovelling. The hand holding the handle is the fulcrum, the other hand provides the effort. With a short movement of the hand, the earth can be moved a long way.

Tweezers

A pair of tweezers is made of two levers joined at the fulcrum. A small squeeze with the fingers in the middle produces a longer movement at the tips, gripping an object firmly. The load is the resistance of the object.

Fishing-rod

The fulcrum is the end of the rod where it is held by the hand. The other hand provides the effort.

The fish is a light load, but it moves through a long distance compared with the hand providing the effort.

Hammer

When a hammer is used to hit a nail, the fulcrum is the user's wrist. The load is the resistance of the wood to the nail. The hammer head is heavy, which means *gravity* adds to the user's muscle force.

The grand piano – a complex of levers

Many machines use combinations of different levers, or a series of levers which act on each other. The piano is a good example.

Hitting a piano *key* produces a sound by making a padded hammer strike a string. The series of levers in between is called 'the action'. It makes the hammer strike the string with a bigger movement than the finger made on the key.

The arrangement allows the pianist to play notes quietly or loudly, and quickly or slowly. 'Piano' is short for 'piano-forte', which means 'soft-loud'.

Inside a grand piano is a long row of hammers – one for each note.

The piano action

When the key is pressed, it raises the jack and roller and the hammer flies up to hit the string.

At the same time, the key raises the padded damper off the string so that the sound can ring out.

If the key is held down, the damper remains away from the string and the sound goes on longer.

The hammer bounces back straightaway and is ready to be played again, even if the key is still half pressed down.

Look for the fulcrums of all the levers involved.

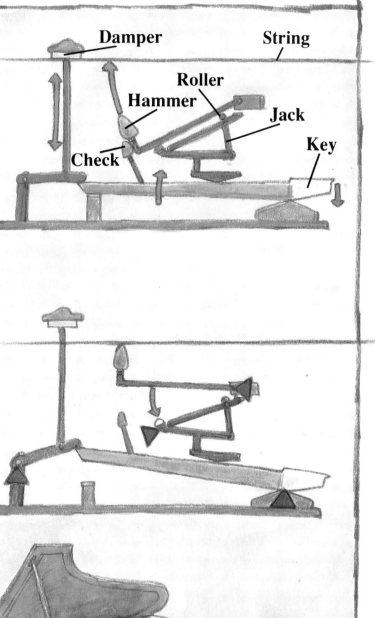

21

More multiple levers

There are many other examples of machines which use *multiple* levers. Here are some of them. You can probably think of more!

Key

Fulcrum Effort Load

Pedal bin
The pedal bin works by turning a small press from the foot into a movement big enough to open the lid.

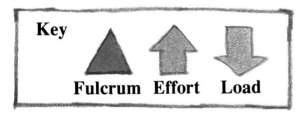

Corkscrew
This type of corkscrew uses levers to make pulling the cork from the bottle a lot easier. The ends of the two levers have teeth, which push the screw up as the levers are pushed down.

Typewriter

A manual typewriter is similar to a piano except that, although the keys are in rows, all the hammers have to strike in the same place, so the hammers are arranged in a semi-circle. Again, a small movement of the finger produces a large movement of the type hammer.

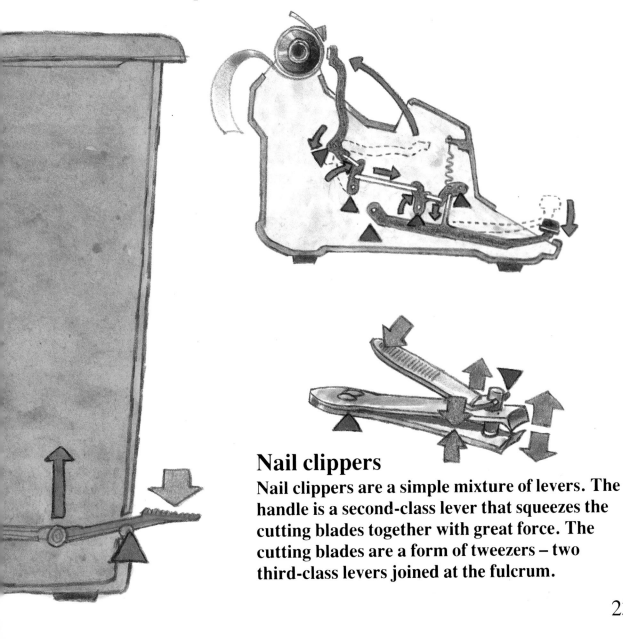

Nail clippers

Nail clippers are a simple mixture of levers. The handle is a second-class lever that squeezes the cutting blades together with great force. The cutting blades are a form of tweezers – two third-class levers joined at the fulcrum.

The weight is in the balance

In 1669, a French mathematician, Gilles de Roberval, invented an arrangement of *parallel* levers now known as the 'Roberval Enigma'. Today it is still used in a simple form of weighing scales. The levers are linked in the shape of a *rectangle*. When one pan moves up or down, the rectangle becomes a *parallelogram*. This allows the pans to remain flat all the time. It does not matter whether the weights are put in the middle of the pans, or at the edges, because the effort or load always appears at the centre. At first sight, this seems to go against the principle of levers. That is why it is called an enigma, which is another word for riddle or puzzle.

If the two weights are the same, no matter where they are on the pans, the scales balance.

Even if one weight is heavier, the pans remain flat.

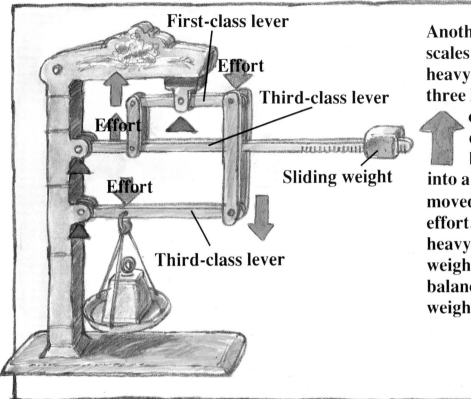

First-class lever

Effort

Third-class lever

Effort

Effort

Third-class lever

Sliding weight

Another kind of scales, used for heavy objects, uses three levers to change a small distance moved by a large load into a large distance moved by a small effort. This way, the heavy object can be weighed, or balanced, by a light weight.

Right *Simple scales are used in markets and shops. These scales work as a first-class lever. The vegetables to be weighed are balanced by a known weight on the other pan.*

Can you find more levers?

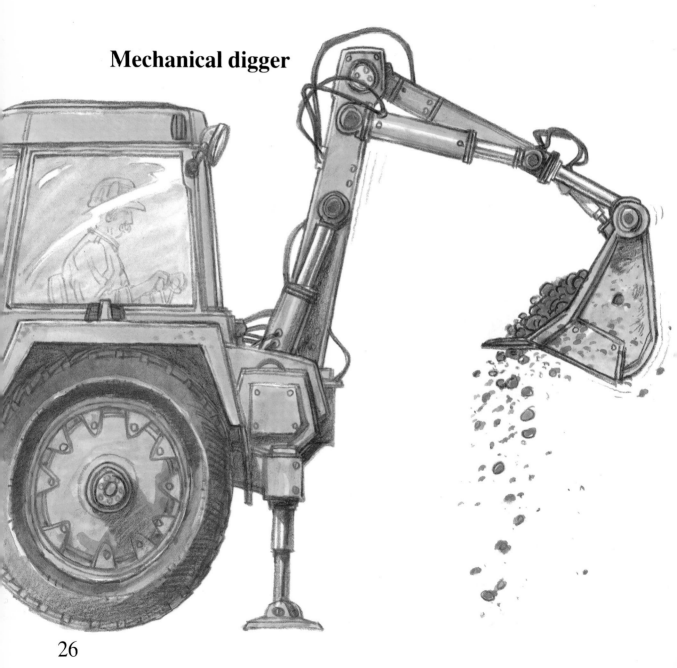

Now you know how levers work and what they can do, have a look around you – at home, in the street, or at school. Can you find any other levers hidden away? These pictures will give you some clues.

Mechanical digger

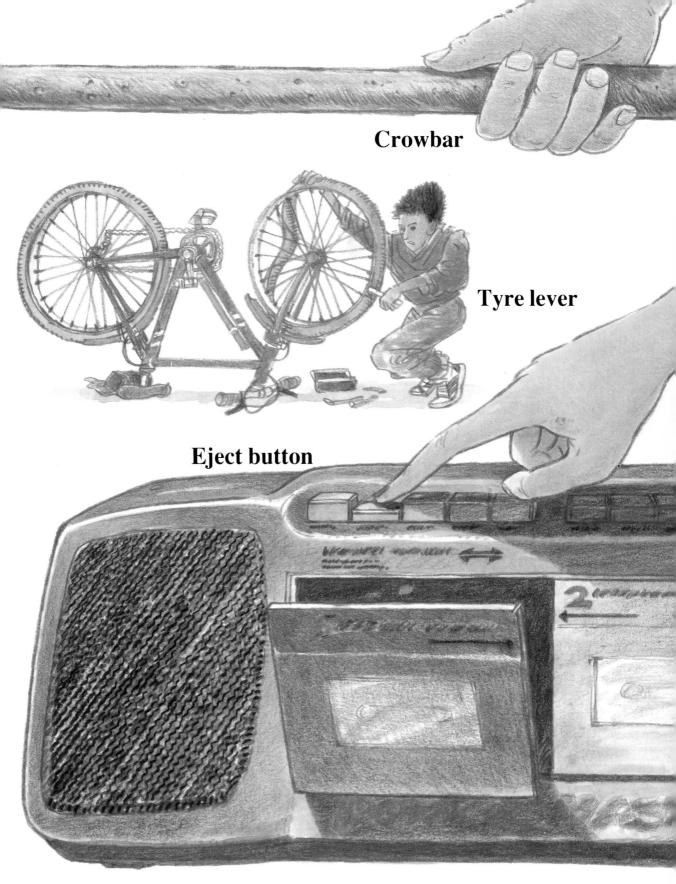

Crowbar

Tyre lever

Eject button

People and levers

People have been using levers since the Stone Age, when they discovered that rocks that they could not lift or push could be moved using levers. We think that Stone Age people used logs as levers to build Stonehenge in England. Stonehenge is a famous ancient circle of enormous standing stones.

Today we use levers in many different ways. We can use them to produce a big force over a small distance, as in nail clippers or tweezers, or to produce a small force over a big distance, as in a golf-club or a fishing-rod. What would the world be like, if people had never discovered the principle of levers?

Levers are used on building sites today. A tower crane is a first-class lever.

Building Stonehenge

Glossary

Balance   Two objects balance each other if they weigh the same. When two people balance on a see-saw neither sinks to the ground.

Blades   Sharp cutting edges, for example on knives or scissors.

Force   The push or pull needed to move objects, or to change their speed or direction. A big force changes the speed of objects more quickly than a small force.

Gravity   The force which pulls objects towards the earth and makes things fall when they are dropped.

Hinged   A hinge is a moveable fastener. Two objects that are hinged are joined together but can swing apart, like a door on a doorframe.

Key   On a piano, one of the white and black levers that are pressed with the fingers.

Machine   A machine is anything made by people to make work easier to do.

Magnified   Made to appear bigger or greater. For example, objects are magnified by a magnifying glass.

Mechanism   A piece of machinery and the bits and pieces inside it. The word can also mean the way a machine works.

Multiple   Having several parts.

Parallel   A word used to describe two lines that are always the same distance apart – like the rails a train runs on, for instance.

Parallelogram   A shape with four sides where each side is parallel to the one opposite it, such as diamonds on playing cards.

Pivot   A point on which something turns freely.

Principles   The laws of science on which the workings of a machine are based. These laws apply to everything in nature.

Rectangle   A shape which is like a square, but longer in one direction than the other. A postcard is a rectangle.

Books to read

For younger readers:
How Machines Work by
Christopher Rawson (Usborne
Publishing, 1988)
How Things Work by Robin
Kerrod (Cherrytree Books, 1988)
Levers and Ramps by Ed Catherall
(Wayland, 1982)
Machines by John Williams
(Wayland, 1990)

For older readers:
Exploring Uses of Energy by Ed
Catherall (Wayland, 1990)
Machines by Mark Lambert and
Alistair Hamilton-MacLaren
(Wayland, 1991)
The Way Things Work by David
Macaulay (Dorling Kindersley,
1988)

Picture acknowledgements

The publishers would like to thank the following for providing the
photographs for this book: Cephas Picture Library 28 (W. Geiersperger);
Chapel Studios 12; Eye Ubiquitous 4 (Paul Seheult), 8 (Paul Seheult),
19, 25; Timothy Woodcock 6; Zefa 20.

Index